79⊱

MW00789223

GERMAN H

BOMBERS

He 177 A-5 (GN+DN) of 5/KG 100 in the summer of 1944.

Do 19 - Fw 200 - He 177 - He 274 - Ju 89 - Ju 290 - Me 264 and others

by Manfred Griehl and Joachim Dressel

Schiffer Military/Aviation History
Atglen, PA

Documentation:

- Aircraft development program from October 1936
- Meeting minutes: Me 261 and Me 264, April 1942
- Performance comparisons of 4- and 6- engine long-range bombers, May 1942
- Quartiermeister Report 7704/44, Me 264, August 1944 – File notes on the meeting of Gen. Qu. 6, Abt., Ju 388 October 1944
- Seekriegsleitung, cessation of He 177 production, June 1944
- General der Aufklärungsflieger, long-range naval reconnaissance in the Atlantic with the He 177, July 1944 – Seekriegsleitung, reconnaissance for U-Boat wartime reconnaissance, October 1944
- Operations report of Luftflotte 3 for the month of July 1943, August 1943
- Seekriegsleitung, Luftwaffe operations in the tonnage battle, June 1942
- Prototype overview of the Fw 200 C, February 1943
- Various flight reports from I/KG 40, 1942-1943
- Prototype testing of the Fw 200 C-5, installation of defensive weapons, October 1942
- File notes on the meeting of Gen. d. Aufklärungsflieger, Ju 290/He 177, July 1944
- Die Seeluftstreitkräfte West im Frühjahr 1940, Report Nr. 40352/40, March 1940

front cover:

He 177s over a convoy

Translated from the German by Don Cox.

Copyright © 1994 by Schiffer Publishing, Ltd.

All rights reserved. No part of this work may be reproduced or used in any forms or by any means – graphic, electronic or mechanical, including photocopying or information storage and retrieval systems – without written permission from the copyright holder.

Printed in The United States of America
ISBN: 0-88740-670-X

This title was originally published under the title, *Deutsche Fernkampfflugzeuge Der Luftwaffe,* by Podzun-Pallas Verlag.

We are interested in hearing from authors with book ideas on related topics.

Literature:

- J. Dressel/M. Griehl, Taktische Miltärflugzeuge in Deutschland 1925 bis heute, Podzun Pallas Verlag, Friedberg 1992
- M. Griehl/J. Dressel, Heinkel He 177-277-274, Motorbuch Verlag, Stuttgart 1990
- B. Lange, Typenhandbuch der Luftfahrttechnik, Bernard und Graefe Verlag, Koblenz 1936
- J. Roeder, Bombenflugzeuge und Aufklärer, Bernard und Graefe Verlag, Koblenz 1990

Photo Sources:

Balke 2
DASA 4
Deutsches Museum 2
Dorneier GmbH 5
Borzutzki 2
Dressel 4
Griehl 3
Heck 2
Heinkel 8
Herwig 3
Junkers AG 5
Lutz 2
Dr. Koos 1
MBB 9
Meier 4
Menke 1
MTU 2
Nowarra 7
Petrick 2
Radinger 2
Ricco 1
Sänger 1
Selinger 5
Siemens AG 3
Stapfer 4
USAF 1
Weber 1

Published by Schiffer Publishing Ltd.
77 Lower Valley Road
Atglen, PA 19310
Please write for a free catalog.
This book may be purchased from the publisher.
Please include $2.95 postage.
Try your bookstore first.

Beginnings in the First World War

In Germany, the development of twin-engine large-scale aircraft began towards the end of 1914. Initially considered "Kampfflugzeuge" (lit. battleplanes) – such as the AEG K I (GZ 1), with the Friedrichshafen FF 36 and the Rumpler 5A 15 (G I) the specialized bomber aircraft came into being. Beginning in the early spring of 1915 this category of aircraft type was designated as G type (Gross-Flugzeug). The development and production of these models ran into serious difficulties at first, so that by October of 1915 only 25 had been produced; by February 1916 a total of only 36 examples had been delivered to the Heeresverwaltung. It wasn't until the summer of 1916 that the Kagohl 1 (Kampfgeschwader der obersten Heeresleitung) was fully equipped with the G types (AEG and Gotha).

Although not playing a decisive role in combat, the operations of these G types gave a clear indication of the future possibilities for strategic air warfare. Within just the few years during which the First World War was fought, performance had increased from 150 kW (130 kmh) to 380 kW (180 kmh). Combat range grew from 500 km to over 830 km. Even the bomb load climbed between the years 1915-1918 from 160 kg to 1000 kg, increasing the takeoff weight by some 150%.

Even more significant was the technical development which occurred with the R types (Riesen-Flugzeug, or giant aircraft). In Russia, Igor Sikorsky had built the first practical long-range airplane, the "Ilya Muromets" in 1913. Germany's Heeresverwaltung issued a demand in September 1914 for a plane which could carry a 1000 kg bomb over a distance of 600 km. In addition, there were naturally special requirements set for lifting capability and operational reliability, as well as night and all weather suitability. For the sake of reliability the engines were to be accessible during flight. On the other hand, high speed, climb rate and service ceiling were given lesser importance.

The slow initial start to such development was not only the result of the state of engine power at the time, but could also be expressed in the contemporary thinking of whether such large aircraft would even have a military purpose at all.

Ferdinand Graf von Zeppelin was the first to take up the challenge. In 1914 he initiated the development of the VGO I (at "Versuchsbau Gotha Ost"). This G-Flugzeug was followed at the end of 1915 by the VGO II and in the spring of 1916 by the VGO III (later called the Staaken R III). Beginning in 1914, the Siemens-Schuckert-Werke also worked on development with a few other, less successful companies (AEG, DFW, Daimler, Linke-Hofmann, etc.). The first military operations were conducted on the Eastern Front beginning in 1916 and in Western Europe in September 1917. In view of the small numbers actually employed the military success was quite insignificant, but the technical achievements can be considered pioneering nonetheless.

The VGO I in its initial form, spring 1915.

For example, the Staaken R XIV could already carry a 1000 kg load over a distance of 1300 km. The takeoff weight was 10350 kg. Five Maybach Mb IVa engines gave the aircraft sufficient power to attain a speed of 130 kmh with a service ceiling of 3700 m. With a 42.2 m wingspan and a length of 22.5 m the name "Riesen-Flugzeug" was truly fitting.

By the war's end, Dornier had also begun work on an all-metal design which bore clearly recognizable similarities to the later Do 11. Junkers, too, had an all-metal concept which was strongly reminiscent of the Ju G 32.

However, the Versaille Treaty prevented any of these projects from bearing any real fruit. For the time being all aircraft production was brought to a complete standstill, with a subsequent relaxing of the treaty's conditions being applicable to smaller aircraft types alone. Only later was it possible, if only in secret, to develop and produce multi-engined aircraft. German industry, by transferring its operations to foreign countries, was nevertheless able to accumulate considerable experience in this field. It was only the area of engine development that Germany would lag behind airframe development, a problem which would continue up until 1945.

Left:
A Staaken R VI in the spring of 1918.

Left:
The Riesenflugzeug Staaken R VI 25/16 during its acceptance.

Above:
The Schütte-Lanz
Grossflugzeug.

Right:
The SSW R VIII,
which was
powered by six
Basse und Selve
BuS IVa engines
each developing
300 hp, 1918.

The Strategic "Heavy Bombers"
Dornier Do 19

The Do 19 – a cantilever mid-wing aircraft with retractable landing gear and twin vertical stabilizers – belonged (along with the Ju 89) to the class of aircraft unofficially designated "Fernkampfbomber", or long-range combat bomber. This class was given strong support by the first Generalstabschef der Luftwaffe, Generalleutnant Walter Wever.

This four-engined strategic long-range bomber was projected to carry a bomb load of at least 1500 kg over 2000 km, giving it an all-up weight of 19000 kg and a ceiling of 8000 m. For the first prototype the Bramo 322 B, with a performance of 530 kW, was available for use as a powerplant. Beginning with the second prototype, however, it was expected that the BMW 132 F (with 1020 kW) would be installed. As armament, it was planned to make use of hand operated traversing mounts fitted with MG 15 guns, located both above and below the fuselage.

The unarmed Do 19 V1 (Werk-Nr. 701, designated D-AGAI) took off on its first flight in October of 1936. The RLM, however, temporarily shelved the idea of the Fernkampfbomber in the fall of 1936 and at that time issued a requirement for a heavily armed "Grossflugzeug" with long range capability. It was proposed that the design would make use of four engines, which would be linked in pairs to a single drive unit with each pair turning a single propeller shaft. Heinkel constructed his He 177 based on this proposal. In April 1937 all further work was halted on the Fernkampfbombern. The components for the Do 19 V2 and V3 prototypes were scrapped, but the Do 19 V1 soldiered on until 1938 in service as a transport plane.

Right:
This model drawing of the Dornier Do 19 "Grossbomber" first appeared in the official sales promotion brochure.

Left:
The first, still unarmed prototype of the Do 19 was powered by four Bramo 322 H-2 radial engines.

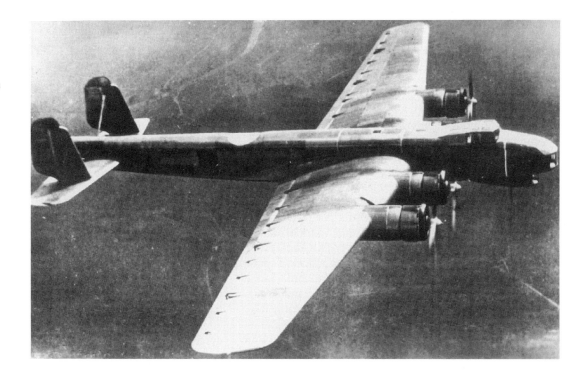

Right:
The first flight of
Do 19 V1 (D-
AGAI) took place
on 28 October
1936 in
Friedrichshafen.

Left:
The cockpit and
forward defensive
position of the
Do 19.

Right:
This air-to-air
photograph of the
planned Do 19
Grossbomber
(V1) was taken
from the upper
rear gun position
of an escorting
Do 17.

Junkers Ju 89

The Ju 89 was designed parallel to the Do 19 within the guidelines of General Wever's recommendation for a Grossbomber. Following Wever's death in 1936, further development of the four-engine strategic bomber was curtailed on the assumption that Germany didn't need a long-range bomber force. It is quite possible that an absence of suitable engines was the primary reason behind the program's discontinuation. In Germany at the time there was no aircraft engine with a minimum performance of 1100 kW, which was the performance needed for a bomber of that size. Eventually the BMW 117, a twelve-cylinder inline engine in the 30 l class, produced a take-off performance of barely 810 kW. At about the same time the Boeing B-17 was under development in the United States –the aircraft which became the most significant bomber with the 8th Air Force in Europe. It is clear that, as a result of inadequate engines the development of a capable bomber aircraft was extremely restricted.

The bomb load and the armament of the four-engined Ju 89 corresponded to that of its competitor, the Do 19, as did its performance figures. The main exception was its takeoff weight, which was significantly higher. The first prototype (Werk-Nr. 4911, designated D-AFIT) of the all-metal low-wing aircraft completed its initial flight in December of 1936 with four Jumo 210 A engines (500 kW) driving Junkers-Hamilton variable-pitch propellers. The Ju 89 V2 (Werk-Nr. 4912, D-ALAT) took off on its maiden flight in the spring of 1937 with four DB 600 A engines (705 kW) powering VDM variable-pitch propellers. This prototype set altitude records on the 4th and 8th of June 1938 with a 5000 kg load to a height of 9312 m and a 10000 kg load to a height of 7242 m, respectively.

The Ju 89 V3 (Werk-Nr. 4913) was modified by Junkers in the summer of 1938 into the Ju 90 large-capacity passenger liner for the Deutsche Lufthansa.

Left:
The unarmed Ju 89 V1 during factory testing in Dessau.

Left:
The Ju 89 V1 had an empty weight of 17015 kg; loaded, the prototype weighed in at 22820 kg.

8

Above: The fuselage of the Ju 89 seen during construction in the main Junkers factory in Dessau.

Below: The armament for the Ju 89 A-1 was to consist of at least two MG 15 and two 20 mm weapons (seen here are the two rear turrets).

Left:
The Ju 89 V1 with modified rudders.

Left:
The unarmed D-AFIT (Ju 89 V1) during factory testing.

Below left:
The fuselage nose of the Ju 89 V1 (Werk-Nr. 4911) during construction.

Below:
The cockpit of the Ju 89 V1 with its bomb aiming device still missing.

The Boeing B-17 (here a later model, which made an emergency landing in Switzerland) was a superior performer in comparison to the Do 19 and Ju 89.

Left: The Ha 142 V2/U1 was reconfigured and tested as an armed long-range aircraft.

At the beginning of 1940 the Ha 142 V2/U1 was utilized as PC+BC with 2/ObdL as a long-range reconnaissance aircraft.

Long-Range Reconnaissance and Auxiliary Bombers
Focke-Wulf Fw 200

Lufthansa's plans for developing a long-range aviation network led in 1936 to the development of the Fw 200 – a 26-passenger high-speed airliner. The cantilever low-wing aircraft, of monocoque construction, was initially powered by four Pratt and Whitney "Hornet" engines, but beginning with the second prototype these were replaced with BMW 132 radial engines. The Fw 200 completed its first flight in July of 1937 under the direction of Kurt Tank.

The RLM showed interest in the Fw 200 – particularly since work on the He 177 was progressing at such a slow pace. Accordingly, Focke-Wulf modified the Fw 200 V10 for military application as a naval reconnaissance aircraft.

The machine was fitted with two aerial cameras for vertical photography, additional fuel tanks in the fuselage and two MG 15s for defensive armament. At the same time, the possibility was explored of using the aircraft to engage naval targets. The RLM accepted this modified airframe, redesignated Fw 200 C, and in September of 1939 issued a contract for a small production batch totaling ten aircraft.

By the summer of 1940 the C-1 series was being delivered with bomb-dropping equipment and MG 15s in the fore and rear upper gunners' compartments, in two of the fuselage windows as well as in the lower gunner's position (with a field-of-fire to the rear and below). In addition, plans were made to install an MG FF in the ventral gondola in a downward forward-firing direction. In terms of ordnance 2100 kg bombs could be carried on the ETCs located beneath the outer engine housings and the inner wing areas, as well as in the bomb bay/ventral gondola. With supplemental installation of fuel tanks the range at a cruising speed of 290 kmh was approximately 4500 km. The maximum speed was 380 kmh at an altitude of 1600 m. Takeoff weight reached 22700 kg, a value which was at the extreme structural limit of the auxiliary bomber. Four BMW 132 H-1 radial engines, each producing 735 kW, were utilized. Aside from its success in attacking Allied shipping the Fw 200 C-1, as well as the structurally similar C-2 (with the exception of its gondola), served in reconnaissance roles and as a convoy shadower working in conjunction with U-boats.

With the Fw 200 C-3 (prototype V13), a structurally improved version was made available using higher-performing BMW Bramo 323 R-2 radial engines (883 kW). The gun positions and the pilot's compartment were armored. The forward ventral station was fitted standard with an MG FF on series-produced models and the upper forward gun compartment was redesigned as a turreted cupola. The outer ETCs could now accept a maximum of two SD 1400s, PC 1700s or SC 1800s. The maximum possible ordnance load increased to 5400 kg for the C-3 and nearly all subsequent versions.

Only a single aircraft (C-3/U2) was produced as a long range reconnaissance version with increased fuel capacity. After fitting non-armored fuselage tanks and drop tanks under the outer engine housings the range was increased to 6400 km. Even the C-3/U3, equipped with the "Atlas-Echelot" in the outer wings, remained a one-off design.

The C-3/U4 version made use of armored auxiliary fuel tanks in the fuselage and had the capability of carrying two S5 aerial torpedoes on the two outer ETCs. The MG 131 replaced the MG 15 in the upper gunner's compartment, the ventral compartment receiving an MG 151/20 at the same time. The crew consisted of seven men.

The Fw 200 C-3/U1 flew armed reconnaissance patrols over the Bay of Biscay and in the northern Atlantic.

The U5 conversion kit included the mounting of an MG 151 in the nose. The Fw 200 C-3/U6 was converted in field shops from the C-3/U4 and, when finished, possessed among other things heavier armament. The modification to the C-3/U7 as a testbed carrier for the Kehl III was not undertaken.

The Fw 200 C-3/U8 had heavier armament and additional fuel tanks in the gondola. Again, the conversion took place at field units.

The C-3/U9 was an example – along with the two C-4/U1 and U2 – of an armed airliner for members of the Reich government.

The "Rostock" surface search radar was installed in the Fw 200 C-4/U3. A short time later this was replaced by the "Hohentwiel" FuG 200 radar system. In particular, Fw 200 C-4 aircraft types were given this modification. Additionally, the armament was beefed up by fitting the HDL 151 with MG 151s in the B1 turret, the D30 in the rear upper gunner's station, an MG 15 in the lower station, the L 151 with MG 151 in the forward gunner's compartment, as well as MG 15s in two side windows.

The maximum speed amounted to 410 kmh at an altitude of 4600 m, with the range up to 4800 km and the service ceiling a maximum of 8400 m. The U4 long-range version was never put into production.

Using the Fw 200 C-5, in 1943 a carrier aircraft for two Hs 293 glide bombs was conceived. This version was also fitted with the FuG 203a "Kehl" and improved armament in the side-firing and lower gunner's positions. With the C-5/U1 the dorsal gunner's position was modified. The U2 was given strengthened armor protection in comparison with the earlier C-4 series.

The improved C-6 series had an FuG 203b transmitter for glide bomb control as well as the FuG 200 surface search radar.

The Fw 200 C-8, similar to the C-4, was developed with an modified gondola to improve the visibility of the bomb controller.

By February of 1944 Focke Wulf had delivered over 260 Fw 200 C combat aircraft. These aircraft bore the brunt of the fighting against Allied ship convoys in the Atlantic. Even Churchill paid these Focke-"Wolves" his respect when he called them the greatest threat to the English sea routes.

Above right:
Werk-Nr. 0256 (an Fw 200 C-8) was fitted with the FuG 200 "Hohentwiel" surface search radar (with a range of up to 80 km).

Right:
An Fw 200 C of I/KG 40 during fueling operations.

Left:
Aircraft of 9/KG 40 in Lecce, Italy in November of 1942.

Right:
Operational machines of KG 40 in western France.

Left:
An operationally ready Fw 200 C-4/U1 (F8+DW) prior to taking off for Rahmel from Prowehren, East Prussia, in August 1943.

Right:
Formation flying as part of an exercise over the Baltic Sea, 1943.

Right:
The forward defensive position, equipped with an MG 131, seen on the Fw 200 C long-range strike and reconnaissance aircraft (Werk-Nr. 360259). (BA)

Left:
A close up photo of the gun position of the Fw 200 fitted with an MG 151/20.

Right:
The rear gunner's position was initially equipped with an MG 15; these were later virtually all replaced with the MG 131. (BA)

Above: Maintenance on the defensive armament of the Fw 200. (BA)

Below: An Fw 200 from KG 40 prepares to take off in Bordeaux-Merignac. (BA)

Right:
Fw Heinz Liekenbröcker, II/
KG 40, in the radio compart-
ment of his Fw 200 C.

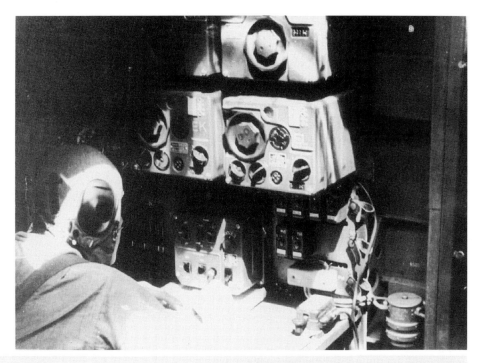

Above:
The Fw 200 C carried ETC racks beneath and next to
the outer engine housings for drop ordnance or fuel
tanks.

Right:
A relatively successfully executed belly landing in an
Fw 200 C-4.

One of the few undamaged Fw 200 Cs captured, April 1945.

Below: Maintenance work at KG 40 in Bordeaux.

Engine run-up tests of the "Sirius" (KG 40). (BA)

The Ju 290, powered by four BMW 801 engines, was the consequential follow-on development of the Ju 90 and served as a large-capacity transport, a long-range reconnaissance aircraft and a long-range bomber. The aircraft was expected – along with the six-engined Ju 390 – to replace the obsolescent Fw 200 over the Atlantic and the Bay of Biscay in the reconnaissance role. Due to the limited number of production aircraft built – less than 50 Ju 290s were actually completed, along with two known Ju 390s – this expectation went unfulfilled. The planned Ju 290 B was to have filled the role as a long-range bomber and guided weapons carrier. Testing for this began in 1943 with the A-5, A-7 and A-8.

Designed by Dipl.-Ing. Bernhard Cruse, the Ju 290 was initially planned as a long-range transport with a large capacity capability. However, the dearth of long-range reconnaissance platforms working in conjunction with submarines led to a request in the summer of 1942 to fit the Ju 290 out as a heavily armed reconnaissance aircraft which would also act as a shadow aircraft against the Allied convoys. To this end the first three Ju 290s, having already been delivered and serving as transport aircraft, were recalled in 1943 and converted over to the long-range reconnaissance mission –along with three additional Ju 290s; the work was completed by late summer 1943. Beginning with the seventh aircraft the Ju 290 was rolling off the assembly line already configured as a long-range reconnaissance aircraft, with a range of over 5000 km.

By the end of 1943 no less than a further twelve Ju 290s were manufactured. Despite this, the manufacturer found himself with a backorder of two aircraft. Three additional Ju 290s were constructed by Junkers as "ultra-long-range reconnaissance aircraft" with an even larger fuel tank system.

By June of 1944 29 Ju 290 As had been built and delivered. Hitler expressed his desire to utilize some of these planes for "ultra" long-range missions. In the plans outlined there was a call for a version with four BMW 801 D engines and a wing surface of 203.6 square meters. By reducing the onboard weaponry the range could be increased by a further 1500 km. There was also discussion of air-to-air refuelling between two Ju 290s in order to conduct reconnaissance flights and strikes along the Trans-Siberian Railway, to make transport flights to Japan and to carry out operations against the supply routes in Africa. The first successful trials with aerial refuelling had already been conducted by Lufthansa in the 1920s. In November of 1943 air-to-air refuelling tests were carried out between a Ju 290 A-4 (Werk-Nr. 110169) and a Ju 290 A-5 (Werk-Nr. 110170).

In addition to Lufthansa – which used the Ju 290 up until April of 1945 on its K 22 flight route (Germany-

The first Ju 290s (here Werk-Nr. 0159) were delivered as long-range transport aircraft and later converted to long-range reconnaissance platforms.

Spain-Portugal) –Fernaufklärungsgruppe FAG 5, Lufttransportstaffel LTS 290 and KG 200 also employed Ju 290 A aircraft. Additionally, FAG 5 also utilized the Ju 290 A-5 to carry three PC 1400 X or Hs 293/294 bombs starting in February 1944.

By the summer of 1944, the experience gleaned in operating the Ju 290 over the Atlantic showed that, even with further development of the passive and active protection, this model would no longer be able to fulfill its assigned tasks. The maximum speed of the Ju 290 A-7 and the new B-series (planned for delivery in the summer of 1944, but never realized) was 450 kmh at optimum altitude; at a range of 5500 km its cruising speed was 330 kmh. The six-engined Ju 390 was to have an estimated maximum speed of 500 kmh over a range of 10000 km, with a cruising speed in the area of 350 kmh. However, these performance figures were considered wholly unacceptable over the continued course of the war. Even with the heaviest armament, airplanes of this class were only able to complete their objectives when flying in large formations. Due to a failure to produce greater numbers of planes, this was not possible.

Therefore – in order to meet the needed ranges for Atlantic reconnaissance missions – efforts would have to be undertaken to increase the speed and altitude performance. The Chief of GL/C-E (Office of Development) accordingly proposed in mid-June of 1944 that the production run of the Ju 290 and Ju 390 be allowed to run out. As early as September of that year all further production work had ceased.

An interim solution was found by utilizing the He 177, which had a higher speed and a range of 5400 km. Manufacture of the Ju 388 and Ju 488 was intended to fill the void in the production capacity freed up by the Ju 290's cancellation. These two designs, by dispensing with overly powerful armament, could reach speeds of 650 kmh. A further increase in speed to over 700 kmh was anticipated by the development branch with the Projekt Hütter 211, which would make use of components from Ju 388 and He 219 production.

Until these measures were realized, plans were underway to produce 18 to 20 Me 264s and use these operationally. A concept which was doomed to failure from the beginning.

The Ju 290 V8 during a long-distance flight.

Left:
The double-tired main landing gear of the Ju 290 A-1 to A-8.

Below:
A Ju 290 A-3 (Werk-Nr. 0161, 9V+DK), which crashed on 26 December 1943.

The Ju 290 A-5 (KR+LA, Werk-Nr. 170) was fitted with the FuG 200 surface search radar.

Below: Three-view of the planned Ju 290 B long-range reconnaissance and bomber airplane (as of 1943).

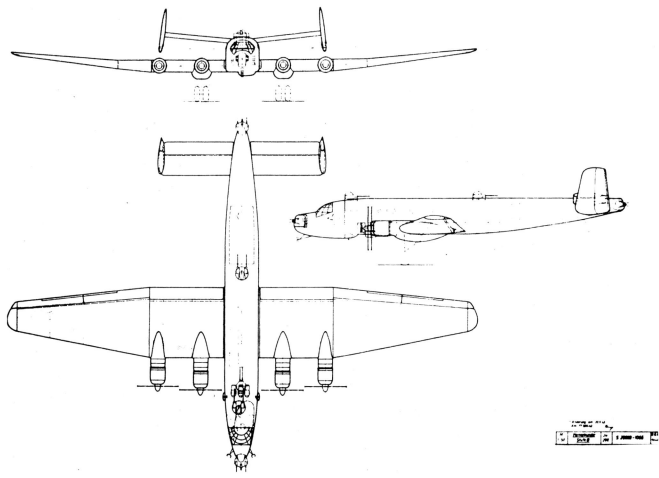

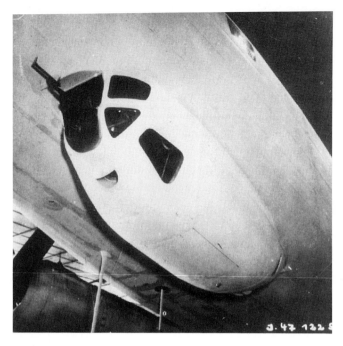

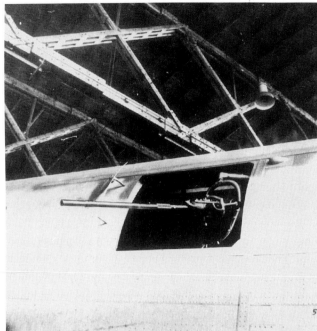

Above:
The as yet unarmed forward defensive position on the Ju 290 A-2 to A-5.

Above right:
The SL 131 side mounts for the Ju 290 A-7.

Right:
Additional fuel tanks in the fuselage of the Ju 290 A.

Below:
Operational aircraft of 1/FAG 5 in Bordeaux-Merignac.

Right:
Maintenance work on an operational machine of FAG 5.

Left:
The 1:1 scale wooden mockup of the forward nose compartment for the planned Ju 290 A-7 differed in a few minor shape details from the version actually utilized.

Right:
Several quad gun arrangements were to be used as the main armament in the planned B through E series.

The Long-Range Bombers
Heinkel He 177

In June 1936 the RLM issued a contract to Blohm und Voss, Heinkel, Henschel, Junkers and Messerschmitt calling for the development of a heavy dive bomber with a 5000 km range and a speed of 500 kmh. Accordingly, Heinkel submitted the Projekt P 1041. The mockup was unveiled in August of 1937 in Rostock-Marienehe. Three months later – after the cockpit area had been considerably reworked – the final inspection took place. At that time the P 1041 (Bomber A) was given the official designation of He 177. Plans initially called for three versions, each with varying ranges (2000, 3000, and 5000 km). Heinkel hoped to make use of four DB 601 A/E engines, coupled together in pairs to form two DB 606 engine units each driving a propeller. In the summer of 1940 Heinkel began production of the individual parts and components.

The He 177 V1 first took off on 11 November 1939. The V2 and V3 followed in the spring of 1940, and by the start of 1941 four additional V-prototypes had joined them. On 24 April 1940 V3 was lost; the same fate befell the He 177 V2 on 27 June 1940.

Beginning in February 1941 the V7 was operated out of Tarnewitz. Along with the V6, it was the first armed version of the He 177.

By the fall of the same year, delays had been encountered in the test program at Rechlin due to bad weather and diverse shortages of DB 606 engines and their casings. The He 177 V8 and the second 0-series model (A-02) were therefore given an enlarged engine housing, which in the long run improved the life of the DB 606.

The first trial flights took place in March of 1942. On the 16th of July A-013 crashed during pitch-and-bank testing in Rechlin. Delivery shortages led to further setbacks in the testing program.

As 1941 turned into 1942 series production of the He 177 A-0 began at Heinkel's Marienehe and Oranienburg plants, as well as at the Arado Werke in Warnemünde.

The He 177 A-014 (Werk-Nr. 0029, GA+QC) for a time belonged to ESt. 177 and was lost during takeoff on 4 July 1942.

After the production of 35 0-series aircraft, Arado undertook license construction of 130 He 177 A-1s at their Warnemünde and Brandenburg sites; the prototype for this series was the He 177 V12. The He 177 A-1 differed from its predecessor by the addition of six 2000 XIIIB bomb racks, giving this version a maximum bomb load capacity of 2200 kg.

Twelve He 177 A-1s were converted to long-range destroyers, designated He 177 A-1/U2, with two MK 101s in the lower nose station, and a few of these modified aircraft were transferred to the planned Fernstzerstörerstaffel of KG 40.

In September 1942 there were renewed problems of structural integrity with the wings, particularly during pitch-and-bank maneuvers and dives. As a result, on the 15th of September Göring ordered that the requirement for the He 177's dive bombing capabilities be dispensed with altogether. In addition, the problems with the engine arrangement hadn't been satisfactorily resolved. Starting in October 1942 the He 177 A-3, with the new DB 610 engines, was to undergo major production at Arado in Brandenburg and Heinkel in Oranienburg. However, since this new engine system had not yet proven operationally reliable and problems had also cropped up with the cooling system and oil circulation, all aircraft delivered up until January 1943 had the older DB 606. At the same time, extensive trials began in Lärz with the first eight He 177 A-3 series aircraft using the DB 610. Due to various difficulties during the test phase, not the least of which was a limited supply of the DB 605 (a pair of which formed the DB 610), severe bottlenecks in production were encountered. The reason was that the majority of DB 605s were needed for fighter production. By the beginning of February 1943 this obstacle had resulted in a shortage of 200 engines. More and more He 177s were therefore put into storage without engines, a situation which would continue up until the war's end. Additionally, the all-up weight of the He 177 had climbed to 34000 kg, necessitating a reduction in the machine's fuel capacity. As a natural consequence, the He 177 A-3/R3 bomber's tactical range dropped significantly.

The He 177 A-3/R3 was planned to incorporate the FuG 203 for operations with up to three Hs 293 guided weapons. The R4 conversion kit, with the "Kehl III" equipment, was made available for installation.

The He 177 V2 seen during a long-range flight over the Baltic Sea. Notice the non-standard armament configuration.

A high-performance long-range bomber became available with the He 177 A-3/R7. As a result of additional fuel tanks, however, only the rear bomb bay could be utilized for dropping ordnance. Later, many He 177 A-1s were converted over to A-3 standards. The planned He 177 A-2 and A-4 high-altitude bomber versions were dropped due inadequate production capacities at the plants.

Production work on the He 177 A-5, also assembled by Arado and Heinkel (for a total of at least 565 examples), began in October of 1943. The A-5 was a direct development of the A-3, with modified armament and a bomb load increased to 2800 kg. The He 177 A-5/R1 also served as a carrier platform for the Hs 293 A-1/B-1 and the PC 1400 X.

A few He 177 A-6 were converted from available A-3 and A-5 airframes. The planned A-6/R1 series made use of a quad-mounted machine gun arrangement for the rear gunner's position, a smaller bomb bay and a fully pressurized nose gunner's compartment. The maximum bomb load had now risen to 3500 kg. The He 177 A-6/R2 was to have a new cockpit shape and a remote-controlled lower nose position with MG 131 Z guns. The upper forward position was fitted with an MG 151 Z, the lower gun position with an MG 131. In the rear gunner's station plans called for a manually-operated HDL 81 mount with four MG 131s.

A small number of He 177 A-5s were converted over to He 177 A-7s, with either DB 610 A/B or DB 613 A/B engines and increased wingspan. Its maximum bomb load capacity was given as 4200 kg. In comparison with the A-6, its armament was to be increased even further.

Heinkel wanted to convert the He 177 A-8, a powerful bomber with four single-unit BMW 801 E engines, from the earlier He 177 A-5. Later, this new aircraft design was assigned the designation He 177 B-5.

The He 177 A-10, too, was to have been equipped with four single-unit BMW 801 E engines. The projected conversion from the He 177 A-7 carried the designation He 177 B-7.

In a decision coming much too late, Heinkel hoped to solve the engine problems once and for all by fitting the He 177 A with four single engines (the He 177 B). The available production capacities, however, did not permit the manufacture of glide bombers and long-range bombers.

The first step in the development came with the He 177 V9. This airframe, drawn from the A-0 series, was converted to become the first (albeit provisional) He 177 B-5 with twin rudders and four single engines. In mid-August 1944 the aircraft was delivered to Rechlin for testing. Other conversions to the B-5 version were undertaken with the V101 (A-3) and the V102 (A-5). Beginning in August of 1944, these aircraft were flying as testbeds with separate engines.

The He 177 V103 (A-5), following its conversion to an "He 177 B-5", was given a manually-operated quad rear gunner's compartment and was available as a test carrier from August of 1944.

The actual prototype for the planned B-5 series, the He 177 V104 (A-5) was not able to be completed by the end of June 1944.

On 25 June 1944, the RLM canceled the planned manufacture of the He 177 B-5 and two weeks later did the same for the entire He 177 production – in favor of production of single-engine fighters. By the time the last He 177 rolled off the assembly line in September of 1944 a total of at least 1140 aircraft had been produced since January 1942.

Maintenance work on an He 177 of I/KG 40 in France during the summer of 1944. The aircraft has a fuselage-mounted ETC for carrying remotely-guided bombs.

Measuring the takeoff and landing distances for the He 177 V5 at a weight of 29000 kg on 18 March 1941.

Below: The He 177 A-1 (Werk-Nr. 15215, VE+UO) crashed on 21 December 1942 during testing at ESt. 177.

Typenblatt

He 177
A-7

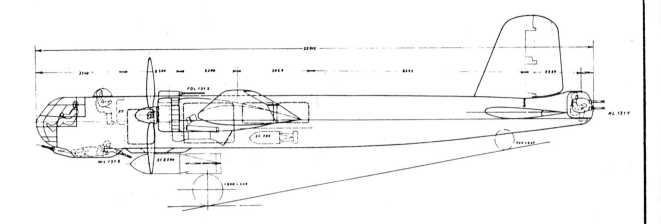

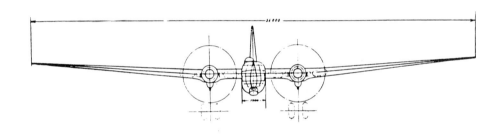

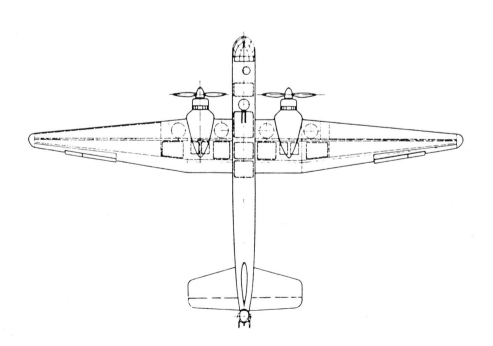

The three-view for the He 177 A-7 was drawn up on 18 August 1943 in Vienna-Schwechat.

Fw Hüppmeier's crew poses at the Minden airfield on 23 April 1944, shortly before taking off for Chateaudun.

Below: An He 177 A-3 of II/KG 40 fitted with the FuG 200.

Loading an He 177 A-5, summer 1944. (BA)

Below: This He 177 A-3 of 2/KG 100 took part in the Operation Steinbock attacks on England and carried the name "Helga." (BA)

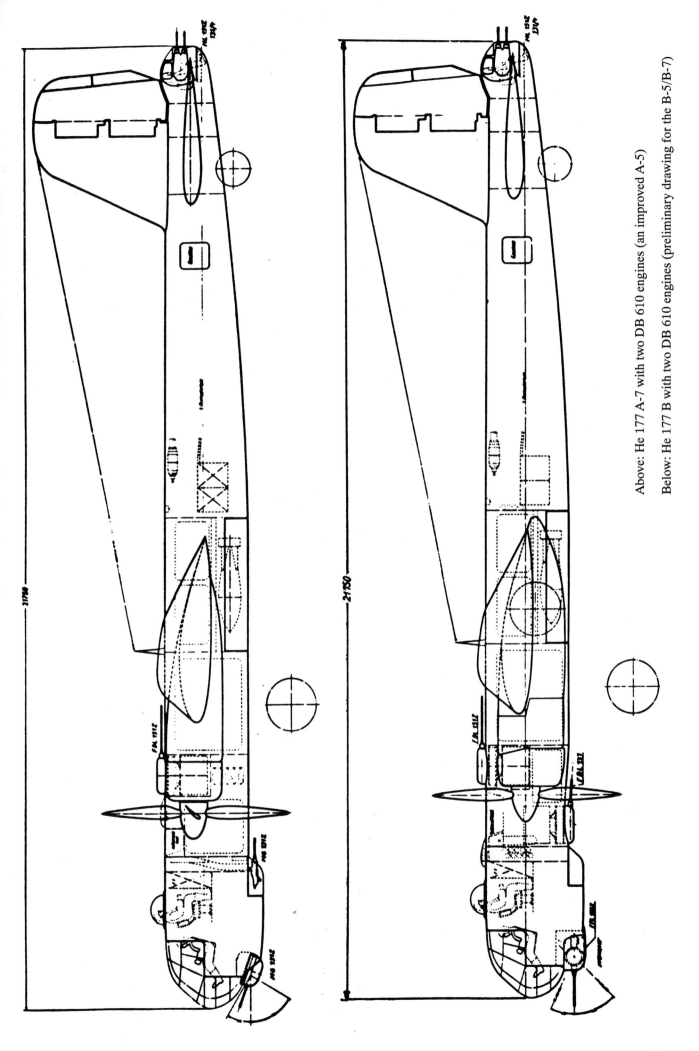

Above: He 177 A-7 with two DB 610 engines (an improved A-5)

Below: He 177 B with two DB 610 engines (preliminary drawing for the B-5/B-7)

Heinkel He 177 H/He 274

Even before the beginning of the war consideration had been given to using the He 177 as a high-altitude bomber. The first rough sketch of a three-place pressurized cockpit was presented by Prof. Heinkel as early as April of 1939. The machine was to be able to operate at an altitude of 15000 m, have a range of 3000 km and carry a bomb load of 2000 kg. In October of 1941 the RLM approved construction of six prototypes of the He 177 H. These were to make use of the fuselage and cockpit area of the He 177 A-1, but would employ four single-unit DB 605 engines with turbo superchargers. Additionally, Heinkel envisioned the use of a twin rudder layout. The RLM recommended that manufacture be assigned in part to French companies in order to relieve some of the burden on the Heinkel plants.

Initially, plans were made to utilize the He 177 V9 to V11 for pressurized cockpit testing, but since the V9 was transferred to hardstand testing only the tenth and eleventh prototypes were available. V10 completed its maiden flight on 21 January 1942, with V11 following a month later. At the same time work was begun on a pressurized, traversing rear turret firing a battery of four guns. Nevertheless, plans for series production of the He 177 A-4 with a pressurized cabin, to begin in the early summer of 1942, were still premature. In early March 1944 both pressurized cockpit testbeds were forced to be mothballed due to a shortage of personnel and space.

In the meantime, significant advances were made in the 1941 developmental contract issued for the He 274. This called for a manufacturing distribution between Heinkel in Marienehe and Farman in Paris.

As weaponry, an MG 131 was to be installed in the lower cockpit area and one MG 131 Z each in the remotely-controlled upper and lower gun stations. Ordnance carried included all available bomb calibers with the exception of single SC 50s, and aerial mines (LMA and LMB), as well as torpedoes. Also planned was the ability to carry remotely-guided weapons, such as the PC 1400 X, the Hs 293 and the Hs 294.

Delivery, planned for the end of July, was delayed due to technical problems with the hydraulics and the sturdiness of the 44 square meter wings, as well as the as-yet unresolved matter of the engines. Individual components were to be constructed in France, then shipped to Heinkel for final assembly. It was only after the beginning of September 1944 that delivery of the first pre-production machine was expected.

On 20 April 1944 the RLM blocked construction of the 0-series. Only the He 274 V1 to V3, plus a fuselage, were to be constructed. Even the planned "safety canopy with ejection seats" was dropped. Shortly prior to this, cancellation of the He 277 was announced, a follow-on development of the He 177 with the four single-unit BMW 801 engines as the He 274.

The Heinkel He 277 V1 (Werk-Nr. 535550, NN+QQ) flew for the first time on 20 December 1943.

In early July 1944 the He 274 V1 was nearly ready for its first flight when Allied forces reached the Farman works. Since it was not possible to transfer the aircraft to Germany, the engines were blown up; the airplane itself remained nearly untouched. The second prototype model was still in an early stage of construction.

At the end of the war the Ateliérs Aéronautiquès des Surenes (A.A.S.), as Farman was called following the company's nationalization, finished construction of the two aircraft under the designations A.A.S. 01 and 02. On 27 December 1945 the A.A.S. 01 (previously He 274 V1) took off on its initial flight and served as a testbed for the S.O. 4000 research aircraft from 1948. The second He 274 flew until 1953 as a test plane with captured DB 603 A engines using TK 11 turbochargers.

Right:
The He 274 V1 long-range bomber, later the A.A.S. 01A, seen in early 1952 in Istres, France.

Below:
Final work on the A.A.S 01. Since DB 603 G engines weren't available, the aircraft was fitted with four DB 603 A-2 engines.

The He 274 V1 in December 1945 in France.

The He 274 V1 on approach in Bétigny-sur-Orge.

Ground testing with the A.A.S. 01 (Atelièrs Aèronautique's des Surenes) as the He 274 A was designated while a French war prize.

The Fi 156 "Storch" parked next to the A.A.S. 01 served as a liaison aircraft when needed.

Below: The A.A.S. 01 was scrapped following trials.

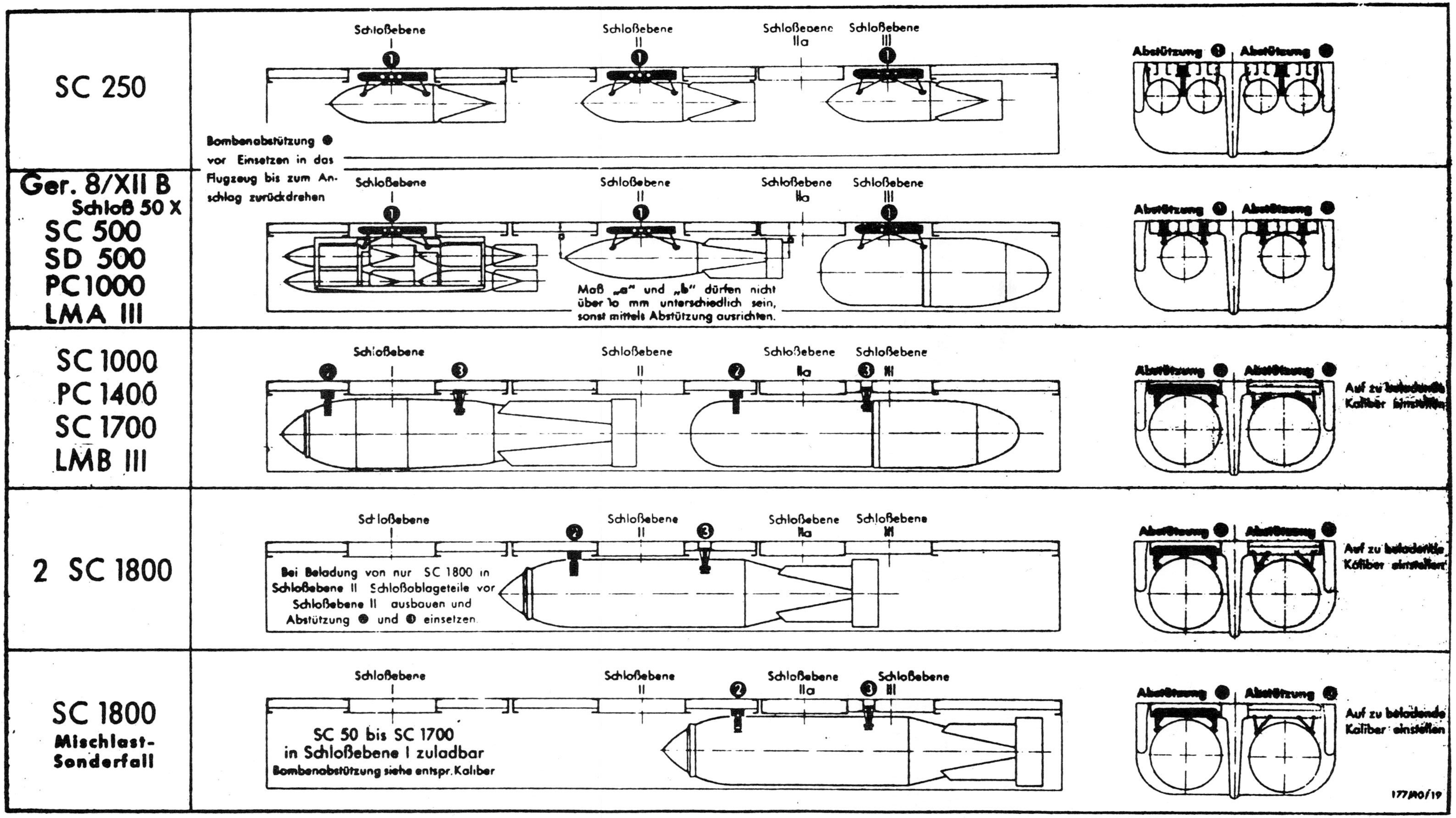

177/140/19

SC 250

Ger. 8/XII B
Schloß 50 x
SC 500
SD 500
PC 1000
LMA III

SC 1000
PC 1400
SC 1700
LMB III

2 SC 1800

SC 1800
Mischlast-
Sonderfall

Abstützung · Ablösung

Auf zu belastende Kaliber einstellen

Schloßebene I
Schloßebene IIa
Schloßebene II
Schloßebene III

Bombenabstützung · vor Einsetzen in das Flugzeug bis zum Anschlag zurückdrehen

Maß „a" und „b" dürfen nicht über 10 mm unterschiedlich sein, sonst mittels Abstützung ausrichten.

Bei Beladung von nur SC 1800 in Schloßebene II Schloßablageteile vor Schloßebene II ausbauen und Abstützung ⊕ und ⊕ einsetzen.

SC 50 bis SC 1700 in Schloßebene I zuladbar
Bombenabstützung siehe entspr. Kaliber

Bomb loading for the He 177 H, showing the SC 250 through SC 1800.

Messerschmitt Me 264

In December of 1942 the unarmed Me 264 V1 completed its maiden flight, powered by four Jumo 211 J engines. Work on this aircraft had started back in 1941. Design data projected a range of approximately 9000 km with a 3000 kg bomb load. Aside from the Me 264 long-range reconnaissance aircraft, with a wingspan of 43 m and a weight of 46000 kg, there were plans for the Me 264 B long-range bomber (with a takeoff weight of 49000 kg) and the specialized reconnaissance platform designated the Me 264 C. The latter had an all-up weight of 56000 kg. Additionally, there was a study completed in early 1942 for a six-engined design.

In the summer of 1944 the Me 264 V1 was lost in an enemy raid. The second prototype was nearly 80 % completed at the time, but was shortly afterward destroyed in an air attack. A further development based on the Me 264 was a reconnaissance model for Atlantic operations, which was expected to replace the Ju 290 and He 177 in the winter of 1945. In an emergency measure, the RLM initially wanted to assemble several Me 264s from presumably available components. However, a report from the end of 1944 showed that these components were nonexistent.

Left:
The unfinished cockpit of the Me 264 V1.

Right:
During the course of testing, the Me 264 was refitted with BMW 801 MA radial engines.

The original design of the Me 264 V1 with four Jumo 211 J-1 engines.

Below: The Me 264 V1 (RE+RN) was, along with the Me 264 V2, destroyed in an air raid and never rebuilt.

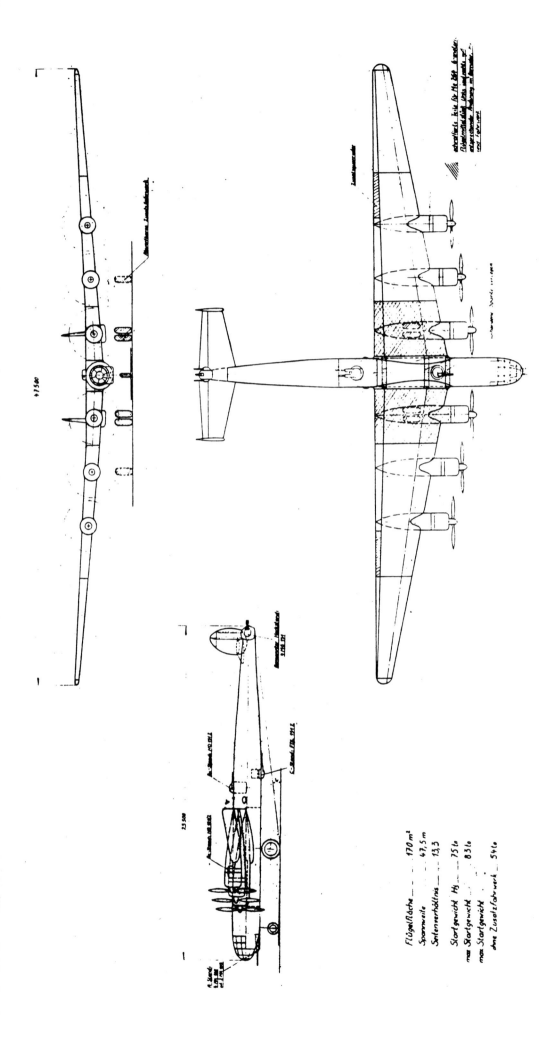

The Me 264 B long-range bomber with six single-unit engines.

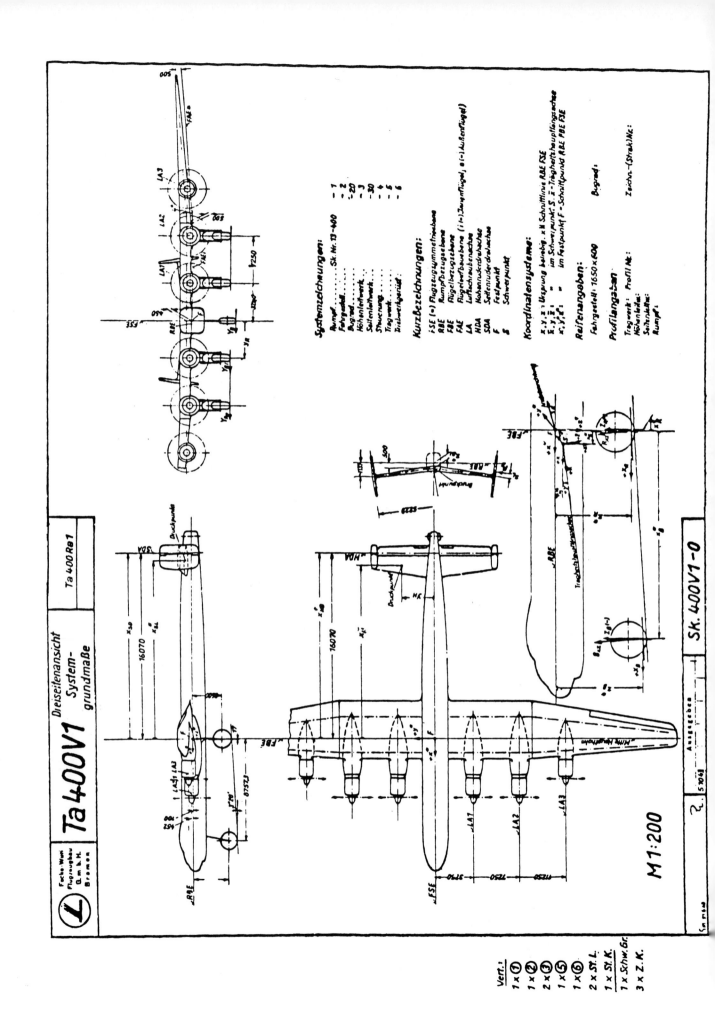

Three-view (basic dimensional) of the planned Ta 400 V1 from 5 October 1943.

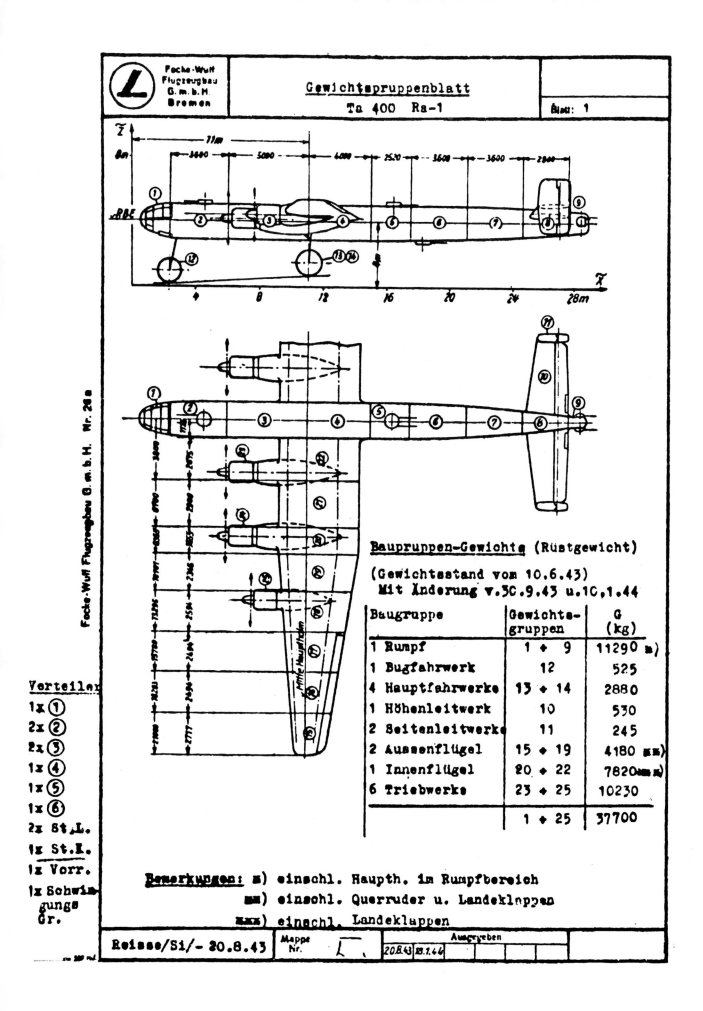

Weight group chart for the Ta 400 Ra-1 (Series 1 construction) from 20 August 1944.

Further designs for large-scale aircraft, such as the Ta 400 for example, remained projects which often never even reached the stage of wind tunnel testing. At this time the Allied air forces ruled the skies over Europe with their four-engined bombers, and over Japan the B-29 "Superfortresses" flew with their deadly cargo. The air powers of the Axis forces couldn't hope to match the performance spectrum and sheer numbers of heavy bombers. This was multiplied by developmental failures in the area of engine production – particularly with jet propulsion engines – and finally, the overestimation of their own resources in raw materials.

The Ta 400 was not built due to the anticipated production of the Me 264.

Below: Officially, two prototypes of the Ju 390 were built, the V1 (GH+UK) and the V2, which was still operating from Lärz in February 1945.

Left:
The Ju EF 100 was to be considered for operations as a long-range aircraft as well as a strategic transport.

Left:
The Sänger'sche Raumgleiter (space glider) was expected to be able to strike any point on the earth without being effectively intercepted. The photograph shows the wind tunnel model which is still in existence today.

Below:
Drawing of the Sänger Raumgleiter "Silbervogel", with a single seat pressurized cockpit, 1944.

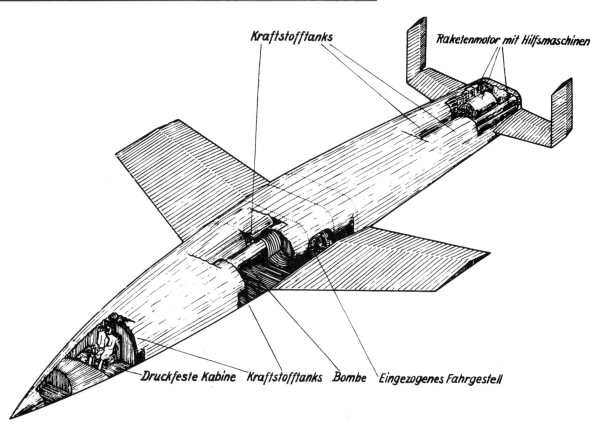

Kraftstofftanks   Rakelenmotor mit Hilfsmaschinen

Druckfeste Kabine   Kraftstofftanks   Bombe   Eingezogenes Fahrgestell

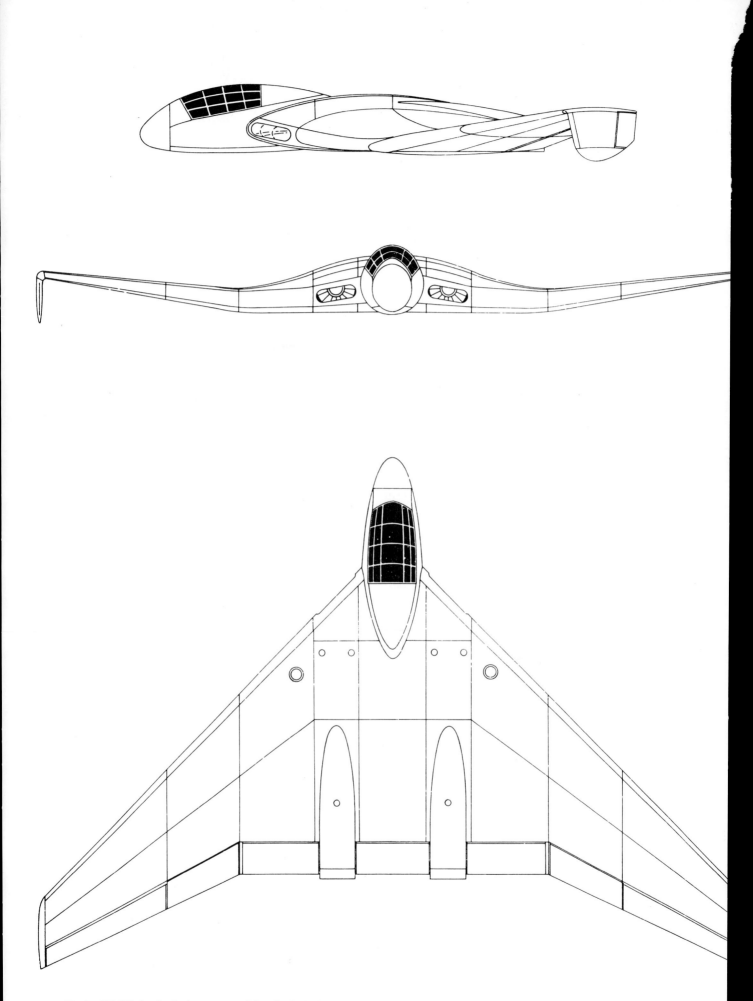

Focke-Wulf's basic design proposal for the "1000x1000x1000" bomber, an aircraft which would have a combat radius of 1000 km and carry a 1000 kg bomb load at a speed of 1000 kmh.

Performance Comparison

Company	Heinkel	Focke-Wulf	Boeing
Model	He 177 A-3	Fw 200 C-4	B-17 F
Crew	5	7	6-10
Engine	2xDB 606	4xBramo 323	4xWright R 1820
Performance(kW)	4000	3000	3600
Width(m)	31.44	32.88	31.61
Length(m)	22.00	23.47	22.77
Wing Surface(m2)	100.0	119.84	131.92
Wing Load(kg/m2)	298.0	189.4	216.0
Weight(kg, empty)	16800	14570	16200
Load(kg)	13000	8130	12300
Total Weight(kg)	29800	22700	28500
Max. Speed(kmh)	480	410	510
Cruise Speed(kmh)	410	305	260
Ceiling(m)	10000	8400	10800
Range(km)	5600	4800	3700
Armament	1xMG 81 I 2xMG 151 4xMG 131	3xMG 151 3xMG 15 1xMG 131	11x.50 cal.
Bomb Load(kg)	2500	5400	5800
Number Produced	>40	>150	<4500
First Flight	1942	1942	1942

Above right: Along with the B-17, the B-24 was the mainstay of the USAAF bombing campaign in Europe.

Right: This B-17 landed in Switzerland during the war; the crew was interred in that country until cessation of hostilities.

Below: This B-24 was painted with the profiles of two additional B-24s to give the impression of a larger formation.

Also from the publisher

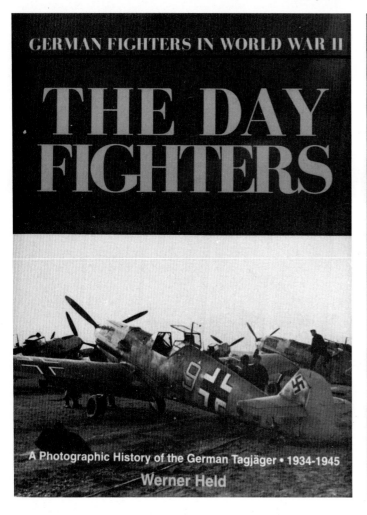

GERMAN FIGHTERS IN WORLD WAR II

THE DAY FIGHTERS

A Photographic History of the German Tagjäger • 1934-1945

Werner Held

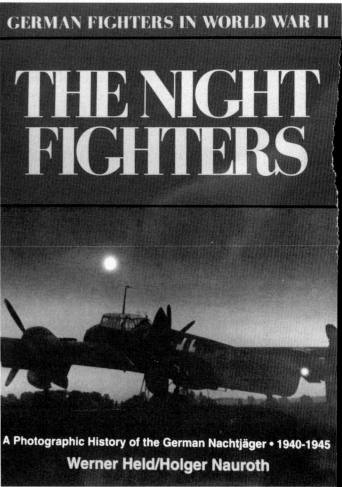

GERMAN FIGHTERS IN WORLD WAR II

THE NIGHT FIGHTERS

A Photographic History of the German Nachtjäger • 1940-1945

Werner Held/Holger Nauroth

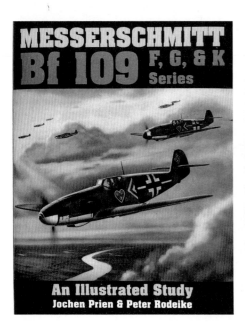

MESSERSCHMITT Bf 109 F, G, & K Series

An Illustrated Study
Jochen Prien & Peter Rodeike

GERMAN AIRCRAFT LANDING GEAR

A Detailed Study of German
World War II Combat Aircraft
Günther Sengfelder

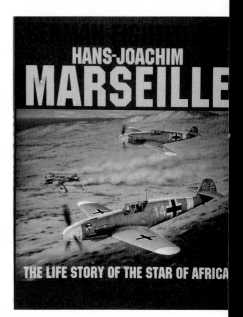

HANS-JOACHIM MARSEILLE

THE LIFE STORY OF THE STAR OF AFRICA

Please write for our free catalog which includes over 200 military and aviation titles.
Schiffer Publishing Ltd., 77 Lower Valley Road, Atglen, PA 19310